DATE DUE			
MAY 28 '94			
JY 31 '02			
AP 25 '03			
JY 0 3 '15			

THE MOON

THE MOON

A SPACEFLIGHT AWAY

by David J. Darling

Illustrated by Jeanette Swofford

DILLON PRESS, INC.   MINNEAPOLIS, MINNESOTA

Photographs are reproduced through the courtesy of the Lick Observatory, the Lunar and Planetary Institute, and the National Aeronautics and Space Administration.

Dillon Press, Inc., 242 Portland Avenue South
Minneapolis, Minnesota 55415

Printed in the United States of America

Library of Congress Cataloging in Publication Data

Darling, David J.
    The moon: a spaceflight away.

    Bibliography: p.
    Includes index.
    Summary: Traces the evolution of knowledge
about the moon, from the invention of the
telescope to the landing of the Apollo astronauts.
    1. Moon—Juvenile literature. [1. Sun] I. Title.
QB582.D37 1984   523.3   84-12644
ISBN 0-87518-262-3 (lib. bdg.)

1 2 3 4 5 6 7 8 9 10 91 90 89 88 87 86 85 84

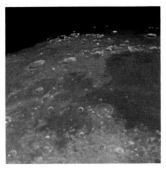

 **Contents**

 **Moon Facts**

Age: More than 4½ billion years

Distance from Earth: *Average*—238,906 miles
(384,400 kilometers)
Closest—221,510 miles
(356,410 kilometers)
Farthest—252,764 miles
(406,697 kilometers)

Distance Across (Diameter): 2,160 miles
(3,476 kilometers)

**Time Taken to Go Around the Earth
(a sidereal month):** 27 days, 7 hours, 43 minutes

Time Taken to Rotate on Its Axis:
27 days, 7 hours, 43 minutes

Temperature: *Day*—more than 212°F (100°C)
Night— -320°F (-195°C)

Gravity at Moon's Surface: About one-sixth that of
the earth

Questions & Answers
About the Moon

Q. If you weighed 100 pounds on the earth, how much would you weigh on the moon?
A. 16½ pounds.

Q. If you can jump 3 feet high on the earth, how high could you jump on the moon?
A. Just over 18 feet.

Q. How many times heavier is the earth than the moon?
A. 81 times.

Q. How many moons would fit inside the earth?
A. 49.

Q. How fast does the moon travel in its path around the earth?
A. 2,287 miles per hour (3,680 kilometers per hour).

Q. Which is the largest crater on the moon?
A. The crater *Bailly*. It is 183 miles (295 kilometers) across.

Q. How long does it take radio signals to travel from the earth to the moon?
A. About 1¼ seconds. (Radio signals travel at the speed of light—186,000 miles per second.)

Q. How many spacecraft have landed or crashed on the moon?
A. At least 37—23 from the United States and 14 from the Soviet Union.

Q. How much did all the moon rocks brought back by the Apollo astronauts weigh?
A. 842 pounds (382 kilograms).

Q. Who was the last person to walk on the moon?
A. *Apollo 17* astronaut, Eugene Cernan.

Q. Is there such a thing as a "moonquake"?
A. Yes. Instruments left on the moon's surface have recorded about three thousand moonquakes per year.

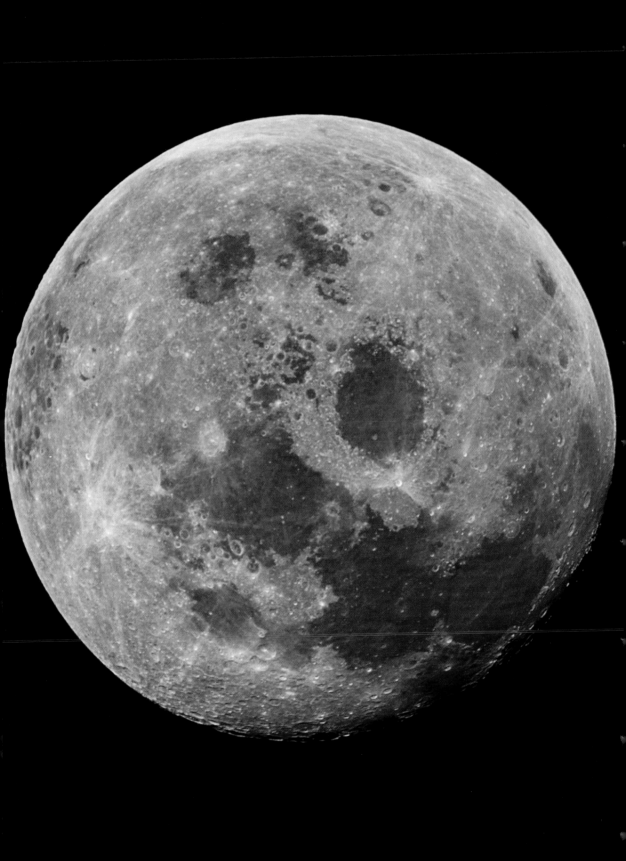

1 Our Spaceflight Begins

Imagine it's the year 2020. From a spaceport near your home town, we've flown in a rocket-powered aircraft hundreds of miles above the earth. Now, inside an orbiting space station, we're waiting for the most exciting part of our journey to begin—a spaceflight to the moon!

Some Facts and Figures

We've tried to learn as much as we can about the world we'll be visiting. The moon is quite a bit smaller than the earth. It's only 2,160 miles (3,476 kilometers) in **diameter,*** compared with earth's 7,928 miles (12,756 kilometers). If the earth were the size of a basketball, the moon would be no bigger than a baseball.

But even though it's small, the moon is special to us. It is the earth's only natural **satellite**—the only object, not made by humans, that orbits, or moves around, our planet. At a distance of 238,900 miles (384,400 kilometers), it's also closer than the sun or the planets. Because it's so close, the moon seems big and bright in our sky.

To get an idea of how close the moon really is, imagine that the whole **solar system** has been shrunk down. In this small scale, the distance from the earth to the sun is only 100 feet (about 30 meters). Now, remember, com-

*Words in **bold type** are explained in the glossary at the end of this book.

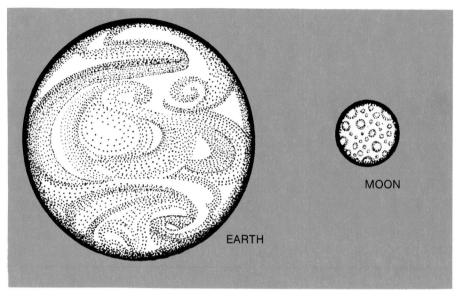

This drawing shows how much larger the earth is than the moon. If the earth were the size of a basketball, the moon would be about the size of a baseball.

pared with most things in space, even the sun is quite close by. But on the same scale, do you know how far from the earth the moon would be? Just 3 inches! The moon is easily our closest neighbor in space.

The moon is not a friendly world, though. It has no water and no air. In fact, it has no **atmosphere** of any kind. During the day, its surface is hot enough to fry an egg. At night, it is colder than the North Pole. We will be visitors to a world where there has never been any life and where everything is strange and new.

Welcome Aboard

Soon our spaceship will be leaving. The captain welcomes us aboard and tells us that, traveling at several miles per second, it will take us less than a day to reach

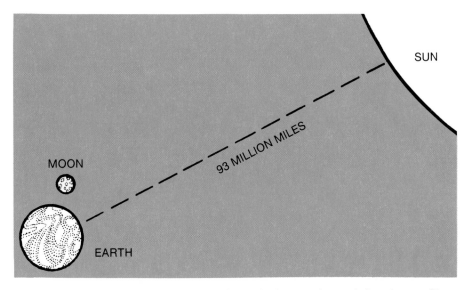

In this picture, you can see that the moon is much closer to the earth than the sun. The sun is 93 million miles away, while the moon, our closest neighbor in space, is about 400 times closer.

the moon. By comparison, he says, the old **Apollo** craft took about four days to make the same journey.

We feel a steady, gentle push as the ship moves away from the space station and gradually gains speed. After about an hour, we reach cruising speed. Lunch is served, and then the captain announces today's in-flight movie, *The Exploration of the Moon.* The cabin lights dim, the film starts to roll, and the sound track begins . . . .

MOON

EARTH

EARTH AND MOON AS PHOTOGRAPHED
BY THE *VOYAGER 1* SPACECRAFT

2 The Moon Seen from Earth

For thousands of years, human beings wondered about the moon. Where did it come from? they asked. What was it made of? What was its surface like? Could life exist there? People wanted to know the answers because the moon was such a big and bright object in the sky.

They were also curious because the moon behaved so oddly. It could be seen both by day and by night. Unlike the stars, it moved across the sky. And, strangest of all, it seemed to change shape! Over each four-week period, it would go from a full circle, to a thinner and thinner **crescent,** to nothing at all, and then back to a full circle again.

Why did the moon behave in this way? Several hundred years ago, scientists began to discover the answers.

The Moon in Motion

Just as the earth orbits the sun, they found, so the moon orbits the earth. Racing at 2,300 miles (3,700 kilometers) per hour, the moon takes 27 1/3 days to go all the way around our planet.

But as the moon circles the earth, it doesn't give off light of its own. It shines by reflecting light from the sun. That is why the moon gradually seems to change shape in

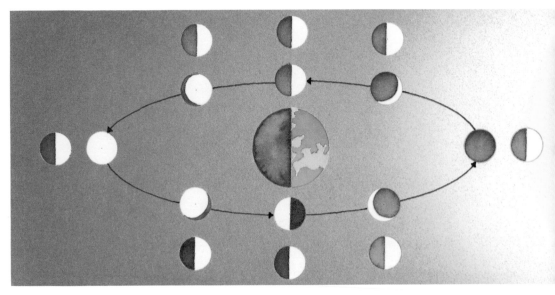

In this drawing, the daylight side of the earth faces the sun. The inner ring of moons shows the moon's phases as seen from earth. The outer ring shows that the moon's

the sky. We see only the moon's side facing the sun.

The side of the moon that faces the sun is lit up. This is the moon's daylight side. When the earth is between the sun and the moon, we see all of the daylight side, and we say that the moon is **full.**

As the moon swings around the earth, we see less of the daylight side and more of the dark side. After a while, all we can see is a thin crescent. Finally, that disappears, too.

When the moon is between the sun and the earth, we can see none of the daylight side. The dark side is turned towards us, and we say that the moon is **new.**

As the moon continues its trip around the earth, its daylight side and crescent shape come back into view. It grows and grows until it gets back to its starting point be-

daylight side always faces the sun.

hind the earth. Then we see the full moon again. All the different shapes of the moon are called its **phases.**

About twice each year, the moon passes exactly between the sun and the earth. Then the moon blocks off all or part of the bright disk of the sun—a **solar eclipse.**

Once in a while, when the moon is on the side of the earth away from the sun, it travels through the earth's shadow. Then, for about an hour, very little sunlight gets through to the moon. This event is called a **lunar eclipse.**

Can you think of anything else that the moon does as it orbits the earth? It spins, like a top, about its own **axis**. We know that it does because the moon always keeps the same side facing the earth. In fact, the time it takes for the moon to do one full spin must be exactly the same as the time it takes for the moon to go once around the earth.

You can do an experiment to see how this spin works. Put a chair in the middle of the room to represent the earth. Now, pretending that you're the moon, walk around the chair sideways so that you are always looking at the chair. By the time you have walked around the chair once, you will also have done one complete turn. And, like the moon, you will have kept your same side facing the chair.

Through the Telescope

People began to figure out the moon's movements, its phases, and its eclipses a long time ago. But they were still puzzled by the moon itself. What, for instance, was the moon's surface like? they wanted to know. Was it covered, in part, by oceans, lakes, and rivers? Was it teeming with

In this photo of the moon, the dark areas are the *maria,* or lowland regions. The lighter areas are the *terrae,* or highland regions. Large craters can be seen covering many parts of the moon's surface.

19

life? Or was it a strange place, unlike anything that had been seen before?

Human eyes alone couldn't provide the answers. But, beginning in 1609, scientists used **telescopes** to get close-up views of the moon's surface. At first, the telescopes were small and simple. Later, they became much larger, and the views they gave became more detailed.

Telescopes showed that, on the side of the moon seen from earth, there are several great dark patches. Some early scientists thought that these were bodies of water. They named them **maria**—the roman word for "seas."

In between the dark maria, telescopes showed much brighter regions. These were believed to be the moon's continents—or **terrae**—that rose above the surrounding seas.

Telescopes today are much larger and more powerful than those used hundreds of years ago. But even the small and simple telescopes first used showed early scientists views they had never seen before.

Most surprising of all, telescopes showed **craters.** These round pits with craggy, raised edges came in all sizes and peppered the moon's surface. They looked like the remains of old volcanoes, and that's exactly what many early scientists thought they were.

Highlands, Lowlands, and Craters Galore

Today, we know that the moon is a very dry world. There is not a drop of water—let alone whole seas—on its surface. We still keep the old names for the moon's dark patches: *Mare Tranquilitatis,* the Sea of Tranquility; *Mare Imbrium,* the Sea of Rains; *Oceanus Procellarum,* the Ocean of Storms, and so on. But we now know that the maria are actually great dust-covered plains. They are flat, low-lying regions of land, not of water.

Craters of many shapes and sizes cover this highland region of the moon. Some craters even have other craters inside them.

In this photo of the *Mare Imbrium,* or Sea of Rains, two large craters and a number of smaller ones stand out on the lowland plain.

Viewed from a lunar orbiter spacecraft, the Apennines mountains on the moon stand out as the lighter colored areas in this lunar region.

The terrae, on the other hand, are highland regions. They contain all the moon's great mountains and mountain ranges, and are much lighter in color than the flat maria. Many of the moon's mountains rise 15,000 feet (about 4,200 meters) or more above the surrounding plains. The highest of all soar more than 25,000 feet (about 7,000 meters), nearly as high as the earth's tallest peaks.

The terrae, or highlands, are also where most of the moon's craters are found. Thousands and thousands of craters can be seen through a large telescope. Some are so small that they would fit inside your living room. Others are 100 miles or more across.

Some craters are very deep and ringed by tall mountains. Others are shallow. Mountain peaks may rise from

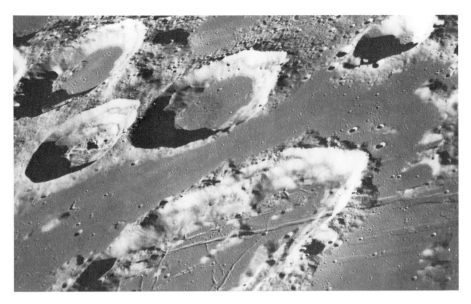

A large crater, *Goclenius,* is shown in the bottom of this photo, taken by the *Apollo 8* spacecraft. Small mountains surround the edge of this crater and the others behind it.

the middle of their flat floors, and bright streaks may spread out for miles in every direction from their edges. In fact, almost every kind of crater exists on the moon.

Scientists used to argue a great deal about how the moon's craters had been formed. There were those who thought that these round, craggy pits were the remains of dead volcanoes. Other scientists believed that the craters were scars made by **meteors** that crashed into the moon's surface a very long time ago.

Who was right? Today, we know that a number of the moon's craters are ancient volcanoes. But most of the craters seem to have been formed by rocks, both big and small, that smashed into the moon in its distant past.

Some craters are found on the smooth floors of the maria. Other strange features are found here, too. There

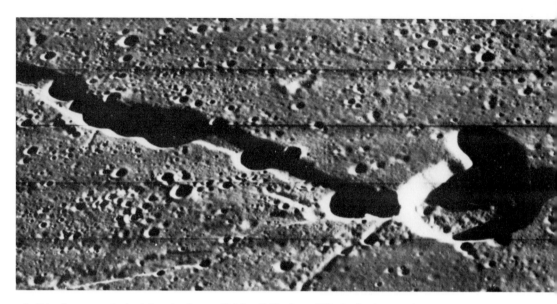

In this close-up photo, taken by *Lunar Orbiter 3,* Hyginus Rille (valley) spreads out to the right and left of Hyginus Crater *(center).* Hyginus Rille is about 6½ miles across

are giant, winding canyons, called **rilles,** that may be hundreds of miles long, several miles wide, and more than a thousand feet deep. There are ridges, as large as 300 feet (about 85 meters) high and 20 miles (32 kilometers) wide, that wind their way across the maria plains. And there are strange **domes**—round bulges up to a few miles across and a few hundred feet high. Some of them have a small crater on top.

Even seen from earth, the moon seems to be a fascinating place. But there's so much that we can't find out using just a telescope. We can't tell what the moon's dust and rocks are made of, and we can't do experiments to try to find out what's inside the moon. Worst of all, we have to guess what is on the moon's "far side"—the side that we can never see from earth.

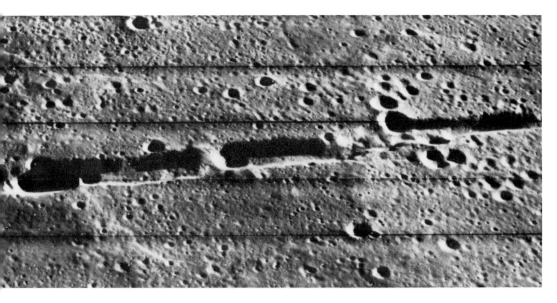

and 2,600 feet deep.

In the 1950s, scientists decided that to learn more about our closest neighbor in space, we would have to explore it with spacecraft. At first, the spacecraft would not carry people. Later, though, they would take astronauts on some of the most exciting missions of all time.

AN ARTIST'S VIEW OF A RANGER SPACECRAFT
JUST BEFORE CRASHING INTO THE MOON

3 Exploring the Moon

In September 1959, the Soviet Union's *Luna 2* crashed into the moon near a big crater called Archimedes. It was the first spacecraft to reach another world. Just a few weeks later, *Luna 3* sped around the moon and sent back pictures of the lunar far side. The exploration of the moon had begun.

The Robot Explorers

Luna 3's pictures surprised everyone. They showed that the moon's far side is quite different from the side we can see from earth. The far side has hardly any low-lying plains. Instead, it's almost totally covered by craters and mountains. Scientists still aren't sure why there should be such a big difference between the moon's two halves.

In April 1962, the United States joined the Soviet Union in exploring the moon. *Ranger 4* became the first American spacecraft to reach the lunar surface. Then, between 1964 and 1965, other Rangers crashed into the moon. As they zoomed in at high speed, they sent back thousands of close-up photographs. These missions were a part of America's plan to land astronauts on the moon.

The Soviet Union's *Luna 9* followed with a lunar soft landing in 1966. Some scientists had thought that the moon might be covered with a thick layer of dust. Such a soft surface would have made it dangerous for astronauts to try to land there. But *Luna 9* didn't sink down. Instead, it sent back the first pictures from the moon's surface.

America's **Surveyor** spacecraft also made several soft landings on the moon. They sent back pictures of cratered, rocky deserts and showed, by testing the soil and rocks, that the moon was a safe place for people to visit.

High above the moon, other spacecraft were carrying out the final plans for America's astronauts to land. These were the **lunar orbiters.** By taking close-up pictures of every part of the moon, they let scientists make the most detailed maps ever of the lunar surface. From these maps, the best places for the Apollo landings could be chosen.

At Cape Canaveral, Florida, scientists check out a Surveyor spacecraft that later made a soft landing on the moon.

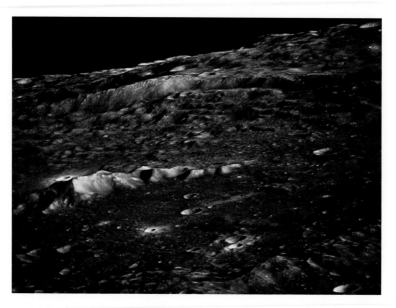

This view of the moon's far side shows that it is rough and mountainous, not a good place for the first Apollo spacecraft to land.

The Human Explorers

On December 21, 1968, many millions of people around the world watched TV pictures of three American astronauts as they orbited the moon in *Apollo 8.* Five months later, the crew of *Apollo 10* took their "lunar module" to within 9 miles (about 14 kilometers) of the moon's surface. Then, came the great event itself.

On July 20, 1969, *Apollo 11*'s Neil Armstrong and Edwin Aldrin made the first human landing on another world. As the two astronauts carefully controlled the flight of their spacecraft, the *Eagle* touched down in the Sea of Tranquility. Later, the astronauts went on a moon walk lasting a little more than two hours. They gathered rock and soil samples, and took hundreds of photographs. Before they left, they set up a number of ex-

Shortly after *Apollo 11* landed in the Sea of Tranquility, astronaut Edwin Aldrin began to set up experiments on the moon's surface. Footsteps made by the astronauts can be seen in the lunar dust.

periments on the lunar surface.

Armstrong and Aldrin were followed, during the next few years, by ten other moon explorers—the astronauts of *Apollo 12, 14, 15, 16,* and *17.* Each moon mission was more difficult and more exciting than the one before. The people of earth saw astronauts standing by giant lunar boulders and rolling stones down the side of a crater. We even saw them driving a special "lunar rover" across the moon's bumpy, dusty surface. But more importantly, because of the Apollo flights, we learned a great deal about what the moon is made of and what may have happened to it in the distant past.

The Unchanging Moon

If we could cut the moon in half, we would find that

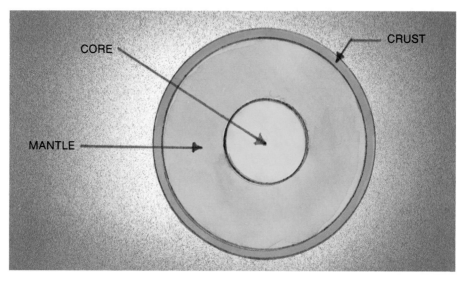

In this drawing, a cross-section of the moon shows that it has the same parts as the earth—a crust, a mantle, and a core.

it's made of three different parts. In the middle is a small ball of hot, heavy, partly melted rocks called the **core.** Around this inner core is a deep layer of lighter, but still partly melted, rocks called the **mantle.** Finally, on top of the mantle is the moon's rocky outer skin, or **crust.**

The earth also has a core, a mantle, and a crust. In the case of the earth, the crust is so thin that it's able to move around on top of the warmer, partly melted mantle. This movement makes the earth's land and oceans change shape over long periods of time. Sometimes gaps open up between different parts of the earth's crust. Then hot **lava** gushes up from the mantle and escapes through volcanoes, or through great cracks, at the surface.

Unlike the earth, the moon's crust today is quite thick and stiff. It doesn't change the moon's appearance by moving around. And it doesn't allow the softer mantle rocks to break through to the surface.

The moon is different from the earth in other ways, too. Since it has no atmosphere, it has no weather, no water, and no rivers or oceans. Without a moving crust, and without rain, wind, and moving water, not much is left that can change the lunar landscape. Only the heat of the day, the chill of the night, and space rocks such as meteors that bump into the surface, can change the way the moon looks.

To scientists, the moon is like a world frozen in time. It is almost the same today as it was billions of years ago. That means many of the rocks brought back by the Apollo astronauts are unchanged from when the moon was quite new. By studying the moon and its rocks today,

The Apollo astronauts brought back many rocks from the moon. By studying moon rocks such as this one, scientists learn much about what happened to the moon long ago.

then, scientists can get a good idea of what has happened to the moon in the distant past.

The Making of the Moon

The moon seems to have formed, like the earth, about 4½ billion years ago. To begin with, it was a very hot ball of liquid rock. Later, its surface cooled and hardened. But exactly how the moon came to be a satellite of the earth isn't known for sure.

The moon may always have gone around the earth. According to this idea, the two objects were formed as close neighbors in space. On the other hand, the moon may once have been a separate planet that was later "captured" as it strayed too near the earth. There's even the chance that the moon was once a part of the earth. It may

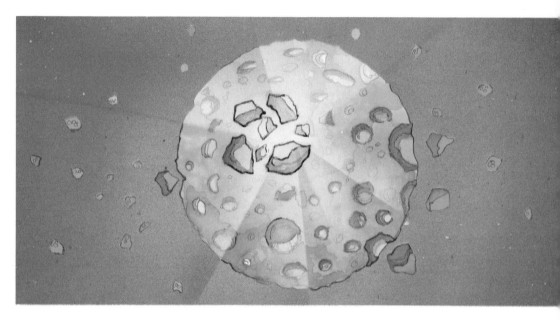

This picture shows how the moon may have looked in its distant past when streaking

have been thrown off, long ago, while our planet was still very hot and rapidly spinning.

However it was formed, afterward the moon began to cool. Its surface hardened into a crust of solid rock. Then a very interesting thing happened. Rocks of every size, from mountains to pebbles, came raining down on the moon's newly-made surface. These were **asteroids** and **meteors** left over from the time when the sun and planets were formed.

Crashing meteors made many of the moon's craters—large and small—that we see today. Giant asteroids dug out the much bigger holes, or **basins,** of the maria.

At the same time, the moon's crust—weakened and softened by the falling meteors and asteroids—allowed hot rocks from the mantle to reach the surface. Volcanoes

meteors and giant asteroids crashed into its surface.

erupted, adding more craters to those made by the meteors.

About 3½ billion years ago, great streams of lava poured through the crust into the maria basins. After a while the maria became filled with lava, a rock that's darker in color than the older, crust rocks of the highlands. Smaller lava flows resulted in such maria features as the strange rilles, ridges, and domes.

About 3 billion years ago, the moon reached the end of its action-packed youth. It looked then very much as it does today. The maria and highlands were formed, the rain of space rocks had stopped, and the crust had once again hardened to prevent the escape of mantle rocks.

For the last 2 or 3 billion years, in fact, very little has happened on the moon. From time to time there has been

About 3½ billion years ago, this rille may have been formed by great lava flows on the moon's surface.

a "moonquake." These quakes have been far less forceful than even a mild earthquake. Scientists have spotted color changes in some craters that may be due to small volcanic eruptions.

At least two large meteors have hit the moon, too. The first, about 1 billion years ago, made the crater Copernicus. The second, perhaps as recently as 100 million years ago, resulted in the crater Tycho. Both Copernicus and Tycho are surrounded by beautiful **rays,** hundreds of miles long, made of light-colored crust material thrown out by the force of the meteor crashes.

From careful studies of the moon rocks brought back by the Apollo astronauts, scientists have learned a great deal about the moon's history. But what have they found out about the rocks themselves?

40

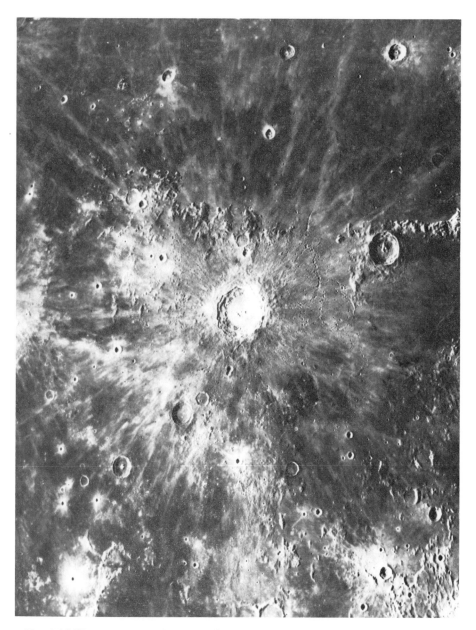

About 1 billion years ago, the crater Copernicus was formed by a large meteor that crashed into the moon. The force of the crash threw out light-colored material that formed the crater's long rays.

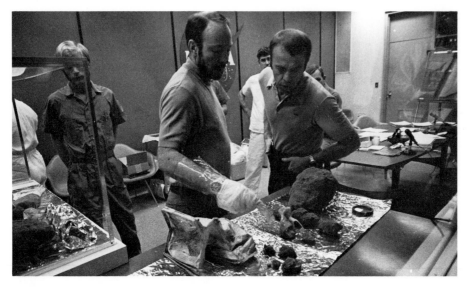

Astronauts Edgar Mitchell *(left, center)* and Alan Shepard *(right)* look at the moon rocks that they brought back from the *Apollo 14* mission. Most of the rocks are basalts.

Pieces of Another World

Most moon rocks are quite like rocks from around volcanoes on earth. They are made of **basalts**—different kinds of crystals that have formed from pools of cooling lava.

Some, taken from the highland parts of the moon, are from 4 to 4½ billion years old. Others, made of so-called KREEP basalts, are slightly less than 4 billion years old. And still others, formed when the great lava flows filled the maria, are about 3½ billion years old.

This last group is made of basalts that have a lot of the metals iron and **titanium** in them. Some people have said that we should try to mine the moon's titanium. In the future, they say, we could use it for building aircraft and spaceships.

This tiny glass "marble" was taken from a moon rock brought back by the *Apollo 15* astronauts. It measures 1.3 centimeters (about half an inch) in diameter.

One of the big surprises of the moon rocks is that they contain tiny glass "marbles." These little balls of glass must have been made from moon dust heated to a very high temperature by meteorites that fell a long time ago.

At first, scientists handled the moon rocks very carefully in case they had any form of life in them. A special place was built in which to study the rocks. That way, if there were strange lunar germs, they couldn't escape to harm anyone. But no signs of life were found. The moon, it seems, is a world that has always been completely dead.

APOLLO ASTRONAUTS GO FOR A DRIVE IN THE LUNAR ROVER

4 A Walk on the Moon

The film has ended. A steward tells us that our spaceship is about to land. Just a few minutes later, dust sprays up past the window, and we feel a gentle bump as our spaceship touches down on the moon's surface.

Spacesuits on, everyone! Since there's no air on the moon, we must take a supply with us. Check your radio. It's the only way we'll be able to talk to each other. Because there's no air, there can be no sound on the moon either.

Check your cooling system, too. Since we'll be on the moon's surface during the day, the temperature outside our spacesuit will be as high as 260°F (127°C). If we were to walk on the moon at night, we'd need a heating system. The temperature then would be bitterly cold— -320°F (-195°C) or even lower!

Now let's step onto the moon. Look at the colors around us. The surface is a light gray, not at all like the bright yellow of the moon seen from earth. The sky is an inky black, studded with stars that look so much brighter than they ever do from beneath earth's blanket of air.

Jump! Even with your spacesuit on, you're several times lighter than you were on earth. The moon's surface **gravity** is just one-sixth that of earth's. Since you weigh less on the moon, you can jump much higher.

Apollo 17 astronaut Harrison Schmitt stands next to a huge, split lunar boulder. Rocks of all sizes and shapes cover the moon.

As we walk around, we can see that the moon's surface is covered with a layer of dust. Notice how fine it is. Every step you take, you leave a perfect print of the sole of your boot.

There are rocks lying everywhere, too. In places, huge boulders stand as big as a house. There are craters, large and small. Some are so worn down that you can hardly tell they are there. Others stand out more clearly. Often the craters overlap or lie inside one another.

In the distance, mountains rise into the inky black sky. Moon mountains have smooth tops and smooth sloping sides. It's hard to tell how far away they are because there's nothing in between—trees or houses or fields—of which we know the size.

Everywhere we look, the land is dead and barren. The

The *Apollo 11* lunar module orbits the moon while the blue, brown, and white earth appears across the distance of space.

moon is certainly not a friendly place to be. But then we notice, in the dark sky above the moon's surface, something that makes us feel much better. It is our home planet, the earth, seen from a quarter of a million miles away—a beautiful blue, brown, and white island of life floating in space.

Appendix A: Discover For Yourself

1. *Explore the Moon With Your Eyes*

The first time the moon was seen through a telescope was in 1609. Before that, people had to use their eyes alone to study our nearest neighbor in space.

Imagine that you're a scientist from 500 years ago. Using just your eyes, try to discover as much as you can about the moon. Can you see the darker and lighter patches? What would you guess them to be? Today, we know that the darker patches are the moon's maria, or lowland plains. The lighter patches are the highlands, which have many craters and mountains.

Draw 29 circles, using a 25-cent piece, in rows on a sheet of paper. Start with the first circle, and draw in the shape of the moon as it appears that day, shading out the part that is dark. The next day, mark the new shape of the moon in the second circle. Do the same on the days that follow. After 29 days, what do you notice? Can you explain the moon's changing shape? Compare your drawings with the picture on pages 16 and 17.

Notice something else, too. The dark, or night, side of the moon isn't completely black. Often,it seems to glow dimly. This glow is "earthshine"—light reflecting onto the moon from the earth.

2. *Explore Close-up With Binoculars*

If you can, borrow a pair of binoculars. These will help you to see the plains, mountains, and even some of the larger craters of the moon, more clearly.

Try spotting the features shown in the photograph on page 19. Or, if you like, try making a sketch of the moon and then comparing it to the photograph. How well did you do? Imagine how difficult it must have been for the early scientists to make maps of the moon using only their eyes, or very small telescopes.

3. Spot a Lunar Eclipse

When the shadow of the earth falls on the moon, there is a lunar eclipse. Below are listed all the lunar eclipses between 1985 and the end of 1999. See if you can spot one!

Eclipses of the Moon 1985-1999

Date	Type	Date	Type
May 4, 1985	Total	June 15, 1992	Partial
Oct 28, 1985	Total	Dec 10, 1992	Total
Apr 24, 1986	Total	Jun 4, 1993	Total
Oct 17, 1986	Total	Nov 29, 1993	Total
Oct 7, 1987	Partial	May 25, 1994	Partial
Aug 27, 1988	Partial	Apr 15, 1995	Partial
Feb 20, 1989	Total	Apr 4, 1996	Total
Aug 17, 1989	Total	Sep 27, 1996	Total
Feb 9, 1990	Total	Mar 24, 1997	Partial
Aug 6, 1990	Partial	Sep 16, 1997	Total
Dec 21, 1991	Partial	Jul 28, 1999	Partial

Appendix B:
Amateur Astronomy Groups
in the United States,
Canada, and Great Britain

For information or resource materials about the subjects covered in this book, contact your local astronomy group, science museum, or planetarium. You may also write to one of the national amateur astronomy groups listed below.

United States

The Astronomical League
Donald Archer,
  Executive Secretary
P.O. Box 12821
Tucson, Arizona 85732

American Association of
  Variable Star Astronomers
187 Concord Avenue
Cambridge, Massachusetts 02138

Great Britain

Junior Astronomical Society
58 Vaughan Gardens
Ilford
Essex IG1 3PD England

British Astronomical Assoc.
Burlington House
Piccadilly
London W1V 0NL England

Canada

The Royal Astronomical Society of Canada
La Société Royale d'Astronomie du Canada
Rosemary Freeman, Executive Secretary
136 Dupont Street
Toronto, Ontario M5R 1V2 Canada

Appendix C:
Apollo Moon Landing Missions

Spacecraft	Astronauts	Landing Date	Landing Place
Apollo 11	Armstrong Aldrin Collins	Jul 20, 1969	Sea of Tranquility
Apollo 12	Conrad Bean Gordon	Nov 19, 1969	Ocean of Storms
Apollo 13	Lovell Haise Swigert	Returned to Earth without landing	—
Apollo 14	Shepard Mitchell Roosa	Jan 31, 1971	Fra Mauro region
Apollo 15	Scott Irwin Worden	Jul 30, 1971	Hadley-Apennines region
Apollo 16	Young Duke Mattingley	Apr 21, 1972	Descartes region
Apollo 17	Cernan Schmidt Evans	Dec 11, 1972	Taurus-Littrow region
			Grand Totals

Total Time Spent Exploring	Distance Covered	Weight of Rocks Gathered
2¼ hours	short	44 pounds (20 kg)
7½ hours	¾ miles (1½ km)	75 pounds (34 kg)
—	—	—
9¼ hours	2 miles (3½ km)	98 pounds (44 kg)
18¼ hours	17½ miles (28 km)	173 pounds (78 kg)
20 hours	16¼ miles (26 km)	215 pounds (98 kg)
22 hours	18 miles (29 km)	237 pounds (112 kg)
79¼ hours	**54½ miles (88 km)**	**842 pounds (382 kg)**

Appendix D:
Important Robot Missions to the Moon

Spacecraft	Country	Launch Date
Luna 1	USSR	Jan 2, 1959
Luna 2	USSR	Sep 12, 1959
Luna 3	USSR	Oct 4, 1959
Ranger 7	USA	Jul 28, 1964
Ranger 8	USA	Feb 17, 1965
Ranger 9	USA	Mar 31, 1965
Luna 9	USSR	Jan 31, 1966
Luna 10	USSR	Mar 31, 1966
Surveyor 1	USA	May 30, 1966
Surveyor 3	USA	Apr 17, 1967
Surveyor 5	USA	Sep 8, 1967
Surveyor 6	USA	Nov 7, 1967
Surveyor 7	USA	Jan 17, 1968
Lunar Orbiter 1	USA	Aug 10, 1966
Lunar Orbiter 2	USA	Nov 7, 1966
Lunar Orbiter 3	USA	Feb 25, 1967
Lunar Orbiter 4	USA	May 4, 1967
Lunar Orbiter 5	USA	Aug 1, 1967
Luna 16	USSR	Sep 12, 1970
Luna 17	USSR	Nov 10, 1970
Luna 20	USSR	Feb 14, 1972
Luna 21	USSR	Jan 8, 1973
Luna 24	USSR	Aug 9, 1976

Description

First fly-by of the moon. Passed within 3,700 miles (5,967 kilometers) of the moon on Jan 4, 1959.

First crash landing on the moon.

First probe to send back pictures of the moon's far side.

A series of crash landers. Each sent back thousands of photographs just before crashing into the moon.

First successful soft lander. Sent back pictures from the moon's surface.

First probe to become a satellite of the moon.

A series of soft landers. Each sent back photographs and carried out soil experiments on the moon's surface.

A series of orbiters. Each sent back thousands of photographs from space that were used to make detailed maps of the moon's surface.

Soft-landed on the moon and then returned to earth with a sample of lunar soil.

Carried the first robot "rover," *Lunokhod 1*, to the moon.

The second successful "sample-return" probe, similar to *Luna 16*.

Carried the second robot rover, *Lunokhod 2*, to the moon.

The last spacecraft to visit the moon (so far!). It was the third successful sample-return probe.

 **Glossary**

Apollo—America's project to land astronauts on the moon. Altogether, six Apollo missions reached the moon, beginning with *Apollo 11* in July 1969, and ending with *Apollo 17* in December 1972

asteroids—large rocks, sometimes several miles across, that go around the sun. About 4 billion years ago, a number of asteroids crashed into the moon and made enormous holes that later filled with lava and became the maria

atmosphere—the layer of gases above a planet's surface

axis—the imaginary line about which a spinning object (such as the sun or the earth) seems to turn

basalt—a type of rock, made of many tiny crystals, that forms from cooling lava

basin—a large, flat area of land that lies below the level of its surroundings

billion—a thousand million. Written as 1,000,000,000

core—the small, heavy, central part of a planet or satellite

crater—a round, bowl-shaped hole made in the surface of a planet or satellite by a volcano or by a falling meteor. Craters are the most common feature on the moon and occur there in nearly all sizes and varieties

crescent moon—a lunar phase in which we can see only a small, curved part of the moon

diameter—the length of a straight line that runs from one side of an object to the other, passing through its center

dome—a rounded bulge, sometimes with a crater on top, formed by hot lava that has pushed up the surface rocks and then hardened. Domes are quite common on the moon's maria

eclipse—the blocking of light from the sun—or any bright object—by something that passes in front of it. See *lunar eclipse* and *solar eclipse*

far side—the side of the moon that is always turned away from the earth. It is covered by craters and mountains and has very few maria

full moon—the lunar phase in which we see all of the moon's daylight side from the earth. It happens when the moon is on the side of the earth farthest from the sun

lava—hot, liquid rock that gushes up to the surface through volcanoes or through other gaps in the crust

luna—Russia's series of unmanned lunar space-craft. *Luna 2* was the first spacecraft ever to land on the moon

lunar—having to do with the moon. Lunar comes from the Roman word, *luna*, meaning moon

lunar eclipse—an eclipse of the moon, which happens when the moon passes through the earth's shadow. The earth then blocks off most of the sunlight from the moon, and the moon appears much darker

lunar orbiter—America's series of unmanned

spacecraft that circled the moon and took detailed photographs of all the moon's surface

mantle—the middle layer of a planet or satellite that lies between the core and the crust. It is usually made of partly melted rocks

maria—the big, flat, dark-colored regions of the moon. *Maria* is the Roman word for *seas*. Today we know that the maria are not seas of water. They are seas of hardened lava that fill huge holes made by asteroids billions of years ago

meteors—small rocks that go around the sun. Meteors that crashed into the lunar surface made many of the craters that we see on the moon today

million—a thousand thousand. Written as 1,000,000

moon—the earth's only natural satellite and our nearest neighbor in space

near side—the side of the moon that is always turned towards the earth

new moon—a lunar phase in which we cannot see the

moon. It happens when the moon is on the side of the earth nearest to the sun

phases—the different shapes of the moon as seen from earth, such as a full moon or a new moon

Ranger—America's first series of unmanned lunar spacecraft. The Ranger cameras took close-up photographs of the moon before crashing into its surface

rays—bright streaks of light-colored dust that stretch out in all directions from some lunar craters

rille—a type of canyon, or steep-sided valley, on the moon. Rilles are found on the maria and were made by lava flows

satellite—an object that circles around a planet. The earth has only one *natural* satellite—the moon—but many *artificial*, or human-made, satellites

solar eclipse—an eclipse of the sun, which happens when the moon passes exactly between the sun and the earth. The moon then blots out all, or part of, the sun as seen from certain places on earth

solar system—the sun plus all the objects that go around it, including: planets, moons, asteroids, meteors, and comets

Surveyor—America's series of unmanned spacecraft that soft-landed on the moon

telescope—an instrument that uses mirrors and lenses to improve our view of distant objects

terrae—the highland regions of the moon, where most of the moon's craters and mountains are found. They appear lighter in color than the maria. *Terrae* is the Roman word for *lands*

titanium—a metal, used, among other things, for making aircraft. It seems to be far more common on the moon than on the earth

⁂ Suggested Reading

Adams, Peter. *Moon, Mars, and Meteorites.* London: HMSO, 1977. May be purchased in the United States from Pendragon House, 220 University Avenue, Palo Alto, CA 94301.
A small, but beautifully illustrated, guide to the moon for older readers. Later sections deal with the planet Mars and with meteorites. (Advanced)

Moore, Patrick. *The Concise Atlas of the Universe.* Chicago: Rand McNally, 1981.
A big book, packed with pictures and information about all aspects of space. The section on the moon contains dozens of close-up photographs of lunar craters, valleys, "seas," mountains, and other features. (All ages)

Worden, Col. Alfred M. *I Want to Know About a Flight to the Moon.* New York: Doubleday, 1974.
Takes you on a trip, aboard an Apollo spacecraft, to our nearest world in space. This book is written by the command module pilot of *Apollo 15.* (Beginner)

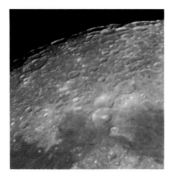

 **Index**